Edith Banks

Nature's Camouflage

ELL BOOKS INC.

Designed and produced by
Albany Books
36 Park Street London W1Y 4DE

First published 1979

Published by Chartwell Books Inc.
A Division of Book Sales Inc.
110 Enterprise Avenue
Secaucus, New Jersey 07094

Printed in Hong Kong

Design: Ivan Cooper

Picture research: Mary Corcoran

Contents

Mimicry of twigs and leaves

The twig in the picture below is not quite what it seems. It is not, in fact, a twig at all, but a caterpillar of the Peppered Moth. These caterpillars, or others like them, are common all over the world, even in urban areas, yet most people have never seen them.

Or rather, they may well have seen them, but never recognized them for what they are because their disguise is so perfect. The purpose of this disguise is not to hide from human beings but to escape the attention of birds. Many birds eat insects and consider caterpillars juicy morsels, but not even the hungriest bird is interested in a twig, and so the twig-caterpillar escapes attack.

This sort of camouflage for survival is common in nature, but there are other uses for camouflage as well. There are hunters whose camouflage allows them to creep up undetected on their prey; even some which look just like their prey so that they can mingle freely with them There are insects which look like flowers to lure other insects to their doom, and flowers which look like insects to get themselves pollinated. It is a strange and fascinating story which we begin with the humble twig-caterpillar.

As can be seen from the photograph, twig-caterpillars look remarkably like twigs. Their bodies are just the right size and shape. Their skins are brown and lumpy, with the sort of irregular knobs and bumps real twigs would have.

The appearance of these caterpillars is only half their camouflage, however. Just as important is the position which the caterpillar maintains, holding itself stiffly away from the branch, and keeping perfectly still. It is capable of retaining this position throughout the day, and only after dark does it move about and feed on the leaves which surround it; at dawn it resumes its disguise.

But do these disguises really work: do

Below: *A twiglike caterpillar of the Peppered Moth. (NSP, P. H. Ward)*

they fool the birds? It is important not to jump to conclusions which seem obvious as these can be completely wrong.

Someone once noticed that a certain type of barnacle was shaped like a goose, and, letting his imagination run away with him, drew a series of diagrams which showed how the barnacles actually turned into geese. It seemed an obvious conclusion to the people of that time and the theory was accepted for centuries. Even now we still call that particular barnacle the Goose Barnacle.

It is unlikely that we would jump to such wild conclusions today. Even so, every idea should be carefully tested. This has been done for the twig-caterpillar's camouflage by scientists under the direction of Professor Niko Tinbergen, who tells the story of one bird's first encounter with them.

The bird was a Jay which had been raised in captivity. It had never had any which were spread out twig-caterpillars it was familiar with twigs. The Jay was put into a large cage on the floor of which were spread out twig-caterpillars and twigs of about the same size. The Jay was used to being fed in this particular cage and so immediately began to search around for food. It hopped about, vainly looking for the expected meal, but paid no attention to the twigs or twig-caterpillars.

Then it stepped on a twig-caterpillar. The caterpillar squirmed and the Jay immediately realized that this was food. It picked it up in its beak, struck it on the ground a few times and ate it. It now hopped across the cage and picked up a real twig. This time it nibbled at the twig, then dropped it, seeing that it was not edible. It tried another twig with the same result. After this the Jay gave up in disgust, and lost interest in both twigs and twig-caterpillars.

This shows that the twig-caterpillar's camouflage *can* work, even though it is unlikely to fool all of the birds all of the time. A bird in the wild is bound to stumble across a twig-caterpillar at some time, and it will then be more likely to find others, because it has acquired what is known as a 'searching image' for them. Similarly it is much easier for us to look for something if we have a picture of it fixed in our minds than if it has been described to us.

Should the bird find another edible twig almost immediately it will be encouraged to keep on pecking at twigs, and, if a number of twig-caterpillars are close by, many of them will be eaten. So twig-caterpillars disperse widely to improve their chances as do other animals which rely on camouflage for survival.

It is important to realize that just fooling some of the birds, some of the time, is enough for evolution to favour these disguises. Evolution works by natural selection, and, in this case, the 'natural selectors' are birds.

Imagine there are twelve caterpillars in a bush. Six look like twigs, and six are just plain brown. During the summer the birds notice four of the plain ones and eat them, but they only notice two of their twig-like companions. The caterpillars which survive become moths and lay eggs, which produce more caterpillars. In this new batch of caterpillars there will be roughly twice as many twig-like caterpillars as plain brown ones, because the offspring take after their parents, and four twig-like caterpillars survived in the previous generation as against two plain ones. If this goes on year after year the twig-caterpillars will eventually predominate.

Evolution has been at work on the animals around us for millions of years now, and the sophisticated masks and disguises which we see among them are the result of the simple process of natural selection endlessly repeated until near-perfection is achieved.

Except that both are insects, the stick-insects, or walking-sticks as they are known in North America, are unrelated to twig-caterpillars, and yet they have evolved a very similar sort of camouflage. They are extremely thin, with spindly limbs, and are coloured green or brown, so that they resemble a mass of sticks or stems.

To augment this disguise they keep as still as possible, holding their front legs straight out ahead and close together, so that these look like a continuation of

Pages 12 & 13: *A New Zealand Prickly Stick-insect. (NSP, P. H. Ward)*

their bodies. Many species freeze completely if disturbed, and even if picked up they remain stiff and motionless, appearing to be dead.

Some stick-insects feed in thorn bushes and have hard prickly skins. As well as providing extra camouflage, these thorns act as a deterrent to any predator which does detect them.

Because of their strange, ungainly shape and large size many stick-insects have lost the power of flight, especially the females, which are usually larger than the males to enable them to produce eggs. For example, the female Violet Stick-insect of eastern Australia has small, stubby wings and cannot fly. The male has beautiful, dark violet wing-membranes, but even he can only glide gently downwards, and if he falls from his tree has to climb the trunk to reach the upper branches again.

The stick-insects belong to an exclusively tropical family of insects, called the Phasmids. This family also includes the leaf-insects. These are found mainly in South-east Asia, but also in Australia, where they are considered rare. Although closely related, they look quite unlike stick-insects, having wide, flattened bodies, instead of thin, spindly ones.

These bodies are intricately marked and coloured to look like glossy, green leaves in some cases; dry, dead leaves in others. The wing veins which are common to all insects have developed in a branched pattern like the veins of a leaf. Even the limbs of these insects may have useless, flat, papery extensions, which look like parts of tattered, half-eaten leaves.

Leaves are everywhere in the insect world, so it is not surprising that many insects resemble leaves for camouflage. Tropical bush-crickets, like the leaf-insects, may mimic dead leaves or green ones, but they have a different angle on

Below: *A leaf-insect which resembles a dry, dead leaf. (NSP, P. H. Ward)*

this disguise, since it is the side-view which is like a leaf, instead of the back.

The green-leaf types of both bush-crickets and leaf-insects do not look perfect, but have one or two small, irregular, brown marks on their otherwise green bodies. These look like the marks left on leaves by disease, or by insects feeding on them, and increase their similarity to real leaves.

Several butterflies and moths look like leaves. The Comma Butterfly in Europe, and its close relative, the Question-mark Butterfly of North Amèrica, resemble dead leaves hanging from a branch. Although the uppersides of the wings are brightly coloured, when the wings are closed up together these vanish completely, and only the dark brown undersides are visible. With their ragged edges they look just like dead leaves, and the small white marks from which the names 'Comma' and 'Question-mark' derive, look like broken spots in the leaf, where the light is showing through.

The Brimstone Butterfly is camouflaged to look like two delicate, yellowish leaves hanging close together, while the richly-coloured, brown Lappet Moth is perfectly disguised as a bunch of dead beech leaves. Both of these hibernate, hanging from twigs throughout the winter, when camouflage is vital as a protection against cold, hungry birds.

Several of the hawk-moths have caterpillars which resemble single, rolled-up leaves. They are green, with diagonal marks like leaf-veins, and have a sharply-pointed tail which looks like the tip of the leaf. But the real triumph of their disguise is the special shading which makes them look flat. This is achieved by a gradual lightening of colour down the sides: from dark green on the upper side to light green on the underside. When sunlight falls on an animal, or any solid object which is

Page 16: *A Comma Butterfly camouflaged amongst dead leaves; the name of this butterfly comes from the small white mark on the wing. (Heather Angel)*

Page 17: *A Dry-leaf Butterfly from Southern Africa. (NHPA, A. Bannister)*

Below: *A bush-cricket which resembles a green leaf; notice the brown spot simulating disease or insect damage to the leaf. (NSP, P. H. Ward)*

evenly coloured, the upperside will look bright because the light hits it directly, but the underside will look darker because it is in the shade. This light-above/dark-below appearance is a clue which we use to tell us that an object is solid, and birds also use it to pick out fat, juicy caterpillars from thin, flat leaves.

In a leaflike caterpillar the device of having the underside light and the upperside dark cancels out this effect. When the sunlight falls on the darker-coloured upperside the caterpillar appears an even shade all over, which makes it seem flat rather than rounded and solid, and greatly increases its resemblance to a leaf. Even on cloudy days this 'countershading', as it is known, still works because the light is falling on the caterpillar only from above.

A simple experiment shows the effectiveness of countershading. Caterpillars which are countershaded are turned upside-down, others are left the right way up, and a count is made of the number eaten by birds. The ones which have been turned upside-down are always consumed in larger numbers because they are so much more noticeable.

The exception which proves the rule is the caterpillar of the Eyed Hawk-moth, which always does hang upside-down when feeding. It is countershaded by having its belly darker and its back light.

It is obviously important that countershaded caterpillars position themselves correctly. The caterpillar does not know about its own camouflage, and has no idea that its life depends on being the right way up, but its instinct tells it to get into the correct position.

These instincts have evolved by natural selection just as the leaflike appearance has, and again the 'natural selectors' are the ever-hungry birds.

Left: *A leaflike Brimstone Butterfly feeds at a flower. (Heather Angel)*

Below: *The Lappet Moth mimics dead beech leaves. (NHPA, L. H. Newman)*

Any caterpillar which did not adopt the best position for disguise purposes would quickly be noticed and eaten, and only those which have the right instinct survive to become moths. When they mate and reproduce they pass on these instincts to their own caterpillars.

Leaf-mimicry is not entirely confined to insects. There is also a fish which, when young, mimics a leaf. This fish is found in the Caribbean, where it lives in lagoons surrounded by mangrove trees. It is coloured yellow with brown patches, matching the dead leaves of the mangroves, and it floats on its side near the surface of the water.

The fish always moves towards dead leaves that are floating on the surface and keeps close to them. If disturbed, it glides slowly down towards the bottom with a gentle zig-zagging motion, just as a leaf would sink. Only when it reaches the bottom does it turn itself the right way up and swim quickly away.

Since twigs and leaves are so abundant they are often imitated, but other devices are used for camouflage. For example, in desert regions, several insects mimic stones. Grasshoppers clinging to grass look like an ear of grass seeds, with their spiky legs and antennae and rectangular shape. Other grasshoppers, and certain mantids, are elongated and coloured to resemble stems of grass.

Many moths resemble broken wood, with a pattern of beige and brown, and an irregular outline. They position themselves on a tree with a fallen branch so that they seem to be part of the splintered stump.

A great variety of creatures, including butterflies, moths, mantids, grass-hoppers, stick-insects and lizards, imitate the lichens which grow on tree-bark. They are usually grey, light-brown, or silvery-green, mottled with white and with wavy, black lines in

Left: *A caterpillar of the Eyed Hawk-moth feeding amongst leaves; 'counter-shading' makes it appear flat and so more leaflike. (Heather Angel)*

Below: *A desert-dwelling grasshopper from South Africa which mimics a stone. (NHPA, A. Bannister)*

Left: *Two insects camouflaged as grass stems: the smaller one is a grasshopper, the larger one a mantid. (NSP, P. H. Ward)*

Below: *A Pale Prominent Moth, one of several which are disguised as broken wood. (NHPA, M. Tweedie)*

imitation of the patterns of lichen. Most of these animals are nocturnal and rest motionless on tree-trunks during the day. Their shape perfects their camouflage in that it does not cast a shadow.

The moths cling very closely to the bark, with their wings spread around them like a cloak, and their antennae tucked away under their wings, out of sight. The edges of the wings touch the bark on all sides, so no shadow is cast. The lizards and mantids are very thin and flattened so that they can lie against the trunk with hardly a bump showing. But they are quite wide across, their bodies tapering outwards into a fine edge, which is pulled in so that it touches the bark when they are at rest. In this way they, too, cast scarcely any shadow.

The Grayling Butterfly cannot solve the problem in this way, since, like all butterflies, it closes its wings up above its body when at rest. As it is unable to flatten itself against the bark, it must cast some shadow, but it minimizes this by positioning itself with its rear-end towards the direction of the sunlight. The sun then hits the edge of the wings and casts only a thin line of shadow.

Several of the moths which are camouflaged as lichen also have bolder lines which mimic the cracks in bark. Some have these going down the wing and these moths always rest head-uppermost on the bark; others have lines across the wings and rest sideways on, so that the wing-markings are aligned with the vertical cracks in the bark.

An experiment was performed with two species of moths which have these stripes to find out how they positioned themselves correctly on the bark. The lines on the bark were reproduced by sticking several strips of black tape onto glass. On one plate of glass the strips were all vertical, on another they were horizontal, to make sure that the moths

did not just orient themselves by gravitational pull, as they could do since cracks in bark are almost always vertical. In fact the moths aligned themselves correctly even if the lines on the glass were horizontal, showing that gravity was not involved. But if a sheet of transparent film was used to cover the strips, allowing the moths to see them but not feel them with their feet, they were unable to position themselves in the right way.

Wherever there are birds there are bird-droppings, and these are a favourite model for caterpillars, especially those of Swallowtail butterflies. Such caterpillars are usually marked with irregular blobs of black-and-white or green-and-white, and some have shiny skins so that they look wet.

Because bird-droppings are found on the tops of leaves these caterpillars can also safely sit in such exposed postions, and theirs is a lazy sort of camouflage, not requiring the odd posture, of, for example, twig-caterpillars. The only problem with this disguise is that it will not work once the caterpillars get too large.

Perhaps for this reason a number of caterpillars which look like bird-droppings when young, change to an inconspicuous green colour as they get older. Immediately the change occurs they take to hiding themselves under leaves instead of sitting on top. It is very tempting to think that they know that they look different and change their tactics accordingly, but this is not so. Their responses are instinctive, and as with their appearance, these instincts have been moulded by natural selection.

Page 26: *A Waved Umber Moth resting on a tree trunk sideways-on; the moths align themselves by feeling the cracks in the bark with their feet. (NSP)*

Page 27: *Caterpillars of the Citrus Swallowtail Butterfly camouflaged as bird droppings. (NSP, P. H. Ward)*

Right: *A Swallow Prominent Moth resting on a tree trunk so that its markings are aligned with those on the bark. (NSP, P. H. Ward)*

Below: *A grasshopper camouflaged against lichen. (NSP, R. Kemp)*

Fading into the background

Pages 28 & 29: *The stripy chick of the Emu is protected by its disruptive coloration. (Bruce Coleman Ltd:, V. Serventy)*

Left: *A green tree-snake blends perfectly with the foliage of a tropical forest. (NSP, P. H. Ward)*

Below: *A Harp Seal pup; like many animals which live near the North Pole it is coloured white to match the snow. (Biofotos, J. M. Terhume)*

Not all camouflaged animals mimic a particular object such as a twig or a leaf. In fact these are a minority: far more common are the animals whose colouring simply permits them to fade into the background and go unnoticed.

Not surprisingly, one of the commonest colours is green. Millions of insects are green. Snakes, lizards, toads and frogs which live in grass or among the foliage of trees are beautiful shades of green, and many forest-dwelling birds, such as parrots, have green plumage. But it is not a colour which mammals can produce in their fur. Only one tree-living mammal does have a greenish camouflage, and this is not of its own doing.

The animal is the Sloth, which lives in the humid South American jungle. It is constantly damp because of the exceptionally high humidity and so the fur is covered with a growth of algae such as live in stagnant water. This gives a greenish tinge to the otherwise grey fur. The sloth hangs upside down and keeps very still as its name suggests, rarely moving. The shaggy, grey-green coat is said to look exactly like an enormous mass of tropical lichen growing on a branch.

Animals which live on tree trunks or branches, or stay close to the soil tend to be brown. In the desert, on the other hand, many animals are a sandy colour, while near the North Pole animals often match the whiteness of the snow. Similarly, in the dimly-lit depths of the ocean most are black or dark red, a colour which looks black in the weak, blue light which filters down through thousands of metres of water.

Countershading is very common among animals which need to be hidden. Most small mammals are countershaded in that their backs are darker than their bellies: the Rabbit is a good example of this. The exceptions are animals such as the Badger and the

Left: *The disruptive coloration of the Oleander Hawk-moth makes it difficult to distinguish from the surrounding leaves when seen from a distance. (NHPA, F. de la Mare-Norris)*

Below: *The caterpillar of the Puss Moth has a strong pattern which breaks up its outline. (Heather Angel)*

Wombat which, being nocturnal, are protected by the darkness, and others like the Skunk which rely on agressive behaviour for their defence.

Almost all fish are countershaded, apart from coral-reef fish and those that live in caves or on the muddy bottom. As well as having a darker back, they often have a silvery belly and partially silvered sides. These silvery surfaces actually reflect light, giving a highly sophisticated sort of countershading. The fish can scarcely be seen when viewed from above or from the side. Only from below would their silhouette give them away.

The strange exception which proves the rule of countershading in fish lives in the River Nile. Known as the Nile Catfish, it has a dark belly and a light-coloured back — the complete reverse of the usual pattern. But it is still counter-shaded, because, for reasons best known to itself, the Nile Catfish swims upside down.

A soldier creeps stealthily through the jungle undergrowth dressed in a combat uniform patterned with patches of green and brown. Far above him an Oleander Hawk-moth rests quietly on a leaf. Unlikely though it may seem the two have a very important thing in common.

Suppose the soldier was wearing a plain brown or plain green uniform. In either case he would be far more easily picked out by the enemy. The patchy colours of his uniform have been care-fully designed to make him as incon-spicuous as possible. Someone scanning the jungle with binoculars in search of

soldiers has his eye out for large, bulky shapes. There is a good chance that he will pass by the soldier who presents the impression of a patchwork of colours, because his size and shape are not so apparent to a casual glance. The patches of different colour break up the outline of the soldier, and he is protected by what is known to biologists as 'disruptive coloration'.

The Oleander Hawk-moth is using just the same trick. Its wings have a marbled pattern of pink and green which, seen close to, is strikingly beautiful. It is hard to understand at first how such coloration could be classified as camouflage at all. But, seen from a distance, the moth is very hard to distinguish from the surrounding leaves, because its outline is broken up by the marbled pattern.

The brightly coloured fish of the coral-reefs often show disruptive coloration, with bold horizontal or vertical stripes. The vivid hues and striking patterns which make them so popular as aquarium fish, are, in fact, a marvellous camouflage against the brilliantly colourful background of coral.

Some fish have gone even further in breaking up their body outlines by having useless extensions, often branched and with tattered or spiky edges, growing out on all sides. The Sargassum Fish, for example, has two spiny protrusions from the top of its head and one from its belly, which look like cacti. These, together with patchy coloration on the body and fins, help to conceal it in the Sargassum weed where it lives.

The Angler Fish rests on the sea-bed by habit, and has tattered strands of skin around its mud-coloured body which make it practically indistinguishable from the mud.

The most extreme example of this disrupted-outline camouflage, is the Sea

Right: *The Sargassum Fish has spiny extensions to the body which help to camouflage it in the Sargassum weed where it lives. (NSP, Marineland of Florida)*

Below: *The bright colours and striking patterns of coral-reef fish act as camouflage against the colourful background of coral. (NSP, L. Perkins)*

Dragon found in Australian waters. This animal is a type of sea-horse with the long spiny body, bony head and tubular snout usually found in such creatures. But the natural outline of its body is entirely obscured by dozens of branched, seaweed-like growths, some even coming from the head and snout.

Many birds have disruptive coloration, especially those which nest among stones. One such is the Ringed Plover, which is patterned in irregularly-shaped bands of black, white and brown across the head and neck. Against a patchwork background of pebbles on a beach, where it sits incubating its eggs, the little bird vanishes.

Like all animals, birds are particularly defenceless when young. It is at this time that camouflage is most important for their survival, as they can neither run nor fly from their attackers, and many species have evolved a disruptive coloration which disappears as they get older. For example, Emu chicks have vertical stripes, while the young of the Woodcock have patterns of curved and straight stripes along the back.

Similarly, some young mammals have protective markings which they lose as they grow older. Deer fawns are usually spotted with white or light beige on a darker background, giving the appearance of dappled sunlight in the forest, while wild piglets are striped in light and dark brown.

Those who cannot flee from predators need camouflage. The female bird who has to sit on the nest relies on her dull brown feathers to protect her. Male birds can afford to be more flamboyant, especially in those species where they mate with the females but have no share in the raising of the chicks. Since one male can mate with several females it will matter little if a large proportion of the males are disposed of by predators.

Where the male and female birds

Below: *A Plover chick on a forest floor; it keeps quite still as the photographer approaches, relying on its camouflage for protection. (NSP, P. H. Ward)*

share the task of incubating the eggs and rearing the chicks, as occurs in over half the bird families of the world, the male is far less flashy and the two sexes tend to look rather similar.

In many species such as the Robin, the male's plumage is a cunning compromise between display and discretion, with bright feathers on the breast to attract a mate, but dull brown or grennish colours on the back and wings. This makes him well camouflaged if seen from above, the vantage point of owls and birds of prey. Attack from below is likely to be by mammals such as foxes and weasels, but most mammals are colour-blind (human beings are a notable exception), so the red breast involves little risk.

In a few species there is a complete reversal of the sex roles. The Dotterel is a wading bird which nests in the inhospitable arctic tundra, where the summer season is very short, and the nest completely exposed because there is no bushy vegetation. If the female were to incubate the eggs only one brood would be possible in the short time available, and, with the high rate of destruction of nest and female by predators, very few young Dotterels would ever be raised.

To overcome this problem the male Dotterel incubates the eggs, while the female wanders off, finds another mate, and lays another batch elsewhere. Thus, if the nest is attacked, the male and one batch of eggs will be destroyed, but all will not be lost. This role-reversal is reflected in the birds' plumage, and also their size, for the male Dotterel is smaller, and of much duller colour and less distinctive pattern than the female. There are other bird species in which the male alone sits on the nest, and here again it is the female which is bright and showy.

Pages 38 & 39: *Many types of wild pigs have stripy piglets; here a pair of European Wild Boar feed with their young in a beech wood. (Bruce Coleman Ltd., Hans Reinhard)*

Page 40 Top: *A female Pheasant sits on the nest, her dull brown feathers and disruptive pattern protecting her from predators. (NSP, P. Burton)*

Page 40 Bottom: *The male Pheasant can afford to be more flamboyant. (NHPA, D. N. Dalton)*

Below: *Unlike adult deer, fawns have a pattern of white patches on the back. (NHPA, J. Good)*

Changing colours in a changing world

Page 41: *Against a background of lichen the normal Peppered Moths are perfectly camouflaged, but the black mutants are very conspicuous. (NHPA, M. Tweedie)*

Below: *A Willow Ptarmigan in July. (NHPA, Jerg Kroener)*

Camouflage is all very well as long as the world about you stays the same colour: but suppose it changes. In a deciduous forest the background colour is green in summer, but becomes brown in autumn as the leaves fall. On the open savannahs the green grass may suddenly turn to brown if there is a drought, or to black if fire sweeps across the plains. In the far north the green-brown landscape is buried under a blanket of pure white snow throughout the winter. The sooty smoke from factory chimneys may change the tree-trunks in a nearby wood from silvery-grey to black. Simply moving about, for some animals, involves changes in the colour and pattern of the background into which they must blend. When the world changes the resourceful change their own colours to match, and the variety of ways in which they do this is truly remarkable.

Insects which live in deciduous trees often have two forms, one green, the other brown, to cope with seasonal changes. The green form is the first generation of the year, which is born in the spring, lives through the summer, lays eggs, and then dies. The young which hatch from these eggs in the autumn are brown, and they live through the winter. In the following spring they lay eggs from which green-coloured young are born.

Grasshoppers are particularly good at colour changes and often need to be. Many live in savannahs which are frequently struck by drought. One species is known to change from green to brown if fed dry food instead of fresh moist food. Obviously, in times of drought the food they eat will be dry, and their environment brown, so if they respond to dry food by turning brown they will be well camouflaged all the time.

Some species of grasshopper can also

turn black, which they do if fires have darkened the savannahs. Others turn pink if they eat large amounts of pink or red food. These species live in regions of tropical forest where they may feed on the red flowers of tropical shrubs. Since they are among these red flowers while feeding they acquire protection by assuming some of their tints. This pink colour will fade in a few days if they stop eating the red food.

Several different mantids which live in savannah regions are also able to change their colour from green to brown when the need arises, but different factors trigger the change. Some respond to a drier atmosphere around them, others to increased amounts of heat and sunlight.

Those which respond to sunlight live in the gnarled, stunted trees which are dotted over the grassy plains of the savannah. These trees periodically lose their leaves but not at any particular time and not necessarily during a drought. The mantids are endangered by being suddenly exposed as green insects on bare, brown trees. When the leaves fall the mantids find themselves bathed in sunlight and respond to this by rapidly turning brown.

Some of the most spectacular changes in colour are seen in animals which live in those parts of the Northern Hemisphere where the ground is covered with snow for several months of the year. The Mountain Hare in Europe, and two hare species in North America, the Prairie Hare, and the aptly-named Varying Hare, gradually moult their brown fur as the weather turns colder and grow a white coat. In spring the process is reversed.

The Ptarmigan is a gamebird which loses the brownish feathers from its back, head and wings and grows white ones. The Arctic Fox turns white, and so does the Stoat, except for the tip of its tail, which remains black, giving us the

Below: *A Willow Ptarmigan in February; the brown feathers have been replaced by white ones to blend in with the winter snow. (NHPA, Jerg Kroener)*

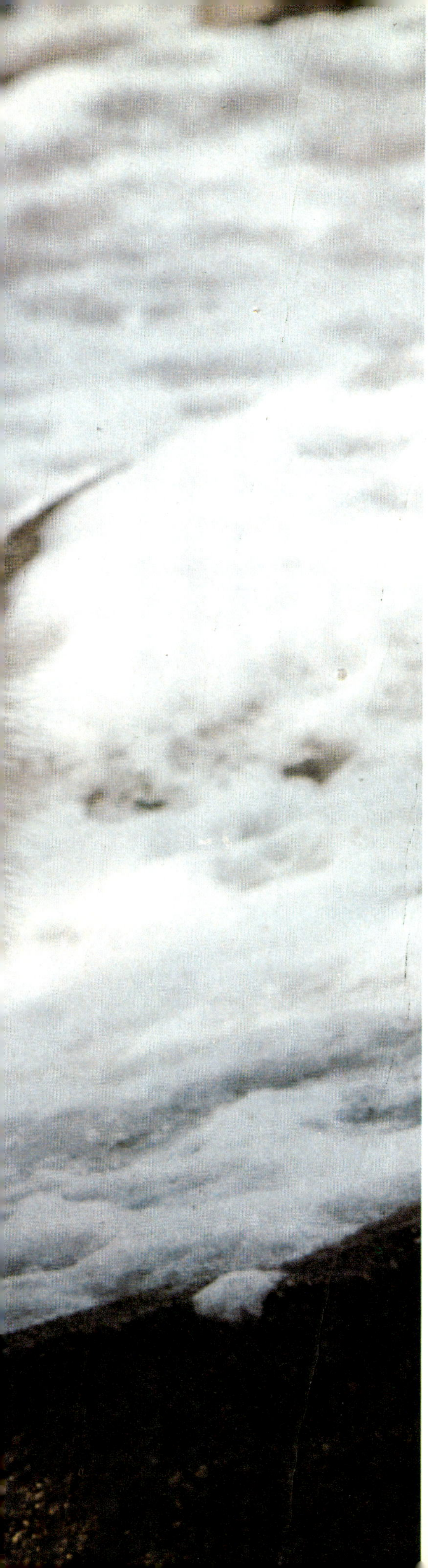

white fur with black, arrow-shaped marks known as ermine.

Colour changes of a very different kind have been caused by pollution. Those most affected have been moths which rest on tree trunks during the day, because the lichens against which they camouflage themselves are killed by the acid smoke which comes from factory chimneys. In woods near industrial areas the lichens are killed, and the tree trunks also blackened by soot. The mottled, silvery-grey, lichen-like pattern of these moths, once so well concealed, is then painfully in evidence.

One such moth which has been intensively studied in Britain is the Peppered Moth. The records of naturalists show that before 1850 a black form of the moth was found only very rarely: at most, one in a hundred moths was black. These black ones were 'mutants': that is, a different type which occurred because of a sudden change (a mutation) in the genes. Genes contain all the information about the structure of an animal and are used as a sort of blueprint as the body develops from a tiny egg into the final, adult form. A genetic mutant is different from its parents and some, but not all, of its offspring will be abnormal too, if it survives long enough to reproduce. But this does not often happen, for mutants are usually severely handicapped by their abnormality. These particular black mutants of the Peppered Moth were very conspicuous against lichen, and quickly eaten up by birds, so they did not become numerous.

With the build-up of industry in the North and Midland regions of England in the nineteenth century, however, the black moths were at an advantage. The original, lichen-like forms became the conspicuous ones and the black ones appeared camouflaged. This resulted in far more black ones than lichen-like ones surviving and with each generation the black ones formed an increasingly large proportion of the total.

The mutant gene spread rapidly so that by 1895 19 out of every 20 Peppered Moths in the Manchester area of north-west England were black. The black form took over in other industrial parts

Left: *An Arctic Fox regrowing its brown coat as spring arrives and the snow melts. (NHPA, T. Stack)*

of Britain and still predominates, whereas in rural south-west England, northern Scotland and west Wales the Peppered Moths remain silvery-grey.

Altogether in Britain there are over 70 species of moth and several other small animals including ladybirds and spiders which have evolved black forms in polluted areas. Similar adaptations have occurred quite independently in industrial regions of Europe, especially the Ruhr Valley in Germany, and in North America.

The mutant ladybirds found in industrial areas are mirror images of the usual form, having red spots on a black background. One study showed that they formed nearly half the ladybird population in a part of Birmingham, England. But after the introduction of anti-pollution laws the black ladybirds decreased in numbers and now form only a quarter of the total.

A very different type of colour change is the rapid transformation which some animals are capable of as they move from place to place. Chameleons are famous for this although their reputation is somewhat exaggerated. Some species can only manage three colours: green, brown and black; others can also change to an off-white shade, yellow and brownish-red. The colour of the skin does not actually change, but appears different. The reason for this is that the pigment is rearranged within the skin with different colours appearing near the surface or concentrated into smaller areas. The pigments are contained within granules which move about in the skin. A complex set of stimuli cause this movement to occur. Light falling directly on the skin effects some changes but the response is mainly what the Chameleon sees in its surroundings.

The Chameleon's colour changes are often unconnected with camouflage. For example, it grows darker when cold. This is probably useful in regulating temperature, since a dark body absorbs more heat from the sun's rays than a light-coloured one. In addition, the Chameleon darkens automatically if angry or afraid, a response which scares predators, especially when combined with the lizard's ability to swell to twice its normal size.

The following story by the naturalist Cyril Crossland demonstrates the usefulness of such colour change: 'While staying in Zanzibar my host's fox terrier showed a hostile interest in a chameleon someone had brought into the house. The chameleon invariably tried to run away when attacked but those who know the species can imagine the ludicrous ineffectiveness of a chameleon's flight. In a few seconds the impossibility of escape seemed to reach the animal's brain, when it at once turned around, opened its great pink mouth in the face of the advancing foe, at the same time rapidly changing colour, becoming almost black. This ruse succeeded every time, the dog turning off at once.'

Far Left: *On a tree-trunk blackened by soot a black mutant Peppered Moth is well concealed. (NSP, J. D. Bradley)*

Left: *A chameleon walking along pine tree branches. (Heather Angel)*

Pages 48 & 49: *The same chameleon a few moments later; as it penetrates deeper into the foliage it becomes greener. (Heather Angel)*

The Decorator Crab attaches green seaweeds to its shell for camouflage. (NSP, Nat Fain)

Improvised camouflage

Some animals are not born with camouflage and have to improvise. Several crabs do this by sticking pieces of seaweed, broken shells, and other debris onto their shells, so that the whole animal is covered by a mask which resembles the sea-bed.

If they move to another part of the sea where the weeds and debris on the bottom are different, they will change the items decorating their shells to match. Seeing a crab do this it is hard to believe that it is not thinking out its improvised disguise. But, like animals which are born with their disguise, it is simply responding to instinct.

Other crabs stick sponges onto their shells which take root and grow there, until eventually they cover the whole shell. There are some small beetles which also have a living camouflage growing on their backs. These beetles, found in the damp rain-forest of New Guinea, have a miniature garden of lichens, mosses and fungi growing all over their backs and legs. The skins of these insects are pitted with small hollows which retain water and give the plants a place to take root and grow.

Stranger still are the caterpillars from Borneo which dress themselves up as part of the flowers on which they feed. Like many caterpillars they have long spines growing all over their bodies, but these are rather stiffer and spikier than most, and for a special purpose. On each spine the caterpillar impales a flower bud. More remarkable still is the fact that the caterpillar, which is not a highly intelligent creature, notices when a flower bud has withered, removes it and replaces it with a fresh one.

Many spiders have improvised camouflage of a different sort. They build two or three dummy 'spiders' in different parts of their web by grouping together the bodies of dead flies. These models are about the same size and shape as the spider, and, with luck, a bird will aim its first peck at one of these rather than at the spider itself. By the time it realizes its mistake the real spider will have fled.

Eyes everywhere

Pages 52 & 53: *Some animals, like this bush-locust, simply display a flash of colour to startle and deter an attacker. (NHPA, A. Bannister)*

Left: *The bright red hindwings of the Red Underwing Moth are covered by the dull forewings as soon as it lands after a flight. (NSP, P. H. Ward)*

Below: *The Javanese Lichen Stick, normally camouflaged as a lichen-covered twig, displays its colourful wings only if alarmed. (NSP, P. H. Ward)*

Even the best camouflage can fail, and when it does so a change of tactics is called for. But an important ingredient in camouflage is to keep still for as long as possible. By the time a predator has got so close that it is obvious the disguise has failed, the animal has usually left it much too late to run successfully for its life. More cunning tactics are needed to get it out of its predicament.

A favourite ploy is to make a sudden move, preferably a gigantic hop or a fast erratic flutter, and then to keep quite still again. These are known as 'dash-and-freeze' tactics. Anyone who has tried to catch a grasshopper or a frog knows how effective these tactics are.

The ploy is even more confusing if there is a display of vivid colour accompanying the movement which vanishes completely when the animal freezes. The pursuing predator is literally dazzled by the colour and finds it far more difficult to locate the animal when it is still and the camouflage is resumed.

For this reason many grasshoppers and locusts such as the Red-winged and Blue-winged Grasshoppers in Europe, and the Yellow-winged Locust in Australia, have showy wings which are only visible when they fly or jump, being tucked away out of sight afterwards. Frogs and toads often have brightly coloured areas on the inside of their thighs which flash when they jump. The Red Underwing Moth, Yellow Underwing Moth and several others have colourful hindwings which, when they are at rest, are covered by their dull, camouflaged forewings. The hindwings make their flight conspicuous, but are lost to view after the moth has landed.

Other animals have refined this technique and display a flash of colour without even moving. It is surprising how effective this can be: most birds are

very easily scared off and other predators may be deterred.

Some of the stick-insects do this by having vivid hues on their wings in sharp contrast to the rest of their body. In most cases these stick-insects have lost the power of flight and displaying to deter attackers is the only function the wings have. One, the Javanese Lichen Stick, is mottled brown and normally camouflaged as a lichen-covered twig. But if disturbed it will open out its short wings giving a dramatic flash of orange, black and white.

Another animal camouflaged as lichen is the Flat-tailed Gecko of Madagascar, which is almost invisible as it rests against tree trunks during the day. If it should be noticed and realizes its danger, the gecko will suddenly open its mouth, the inside of which is garishly coloured in shocking pink and bright orange.

More familiar animals which use this strategy are the many butterflies which have bright uppersides to their wings but dull, well-camouflaged undersides. They generally rest with their wings closed up together, but, if disturbed, will suddenly flash their wings open revealing the colourful uppersides.

Butterflies and moths which flash their wings by way of self-defence very often have circular marks on them which look remarkably like the eyes of much larger animals. Some of these are fairly simple with just a whitish circle enclosing a darker area, as found on the wing-tips of the Peacock Butterfly. Others may be highly sophisticated representations of eyes, with the 'pupils' slightly off-centre giving the impression of an intense, close-range stare, small white marks looking like highlights to the 'pupils', and shading to add a three-dimensional effect. The most impressive eye-spots are found in some tropical species such as the Owl Butterflies of

An Eyed Hawk-moth at rest. Disturbed by the Disturbed by the approach of the photographer, the moth lifts its forewings to display the eye-spots on the hindwings. (NSP, P. H. Ward)

South America whose paired eye-spots look like the staring face of an owl.

Experiments have shown that the appearance of pictures of eyes frighten birds and that the more accurate the pictures are, the more frightened the birds will be. In their natural habitat such eyes would belong to a fox, an owl, a snake or some other enemy, and presumably the birds are geared to escape from them for this reason.

What is interesting is that young birds reared in captivity who have never faced a predator back away from the eye-pictures just as fast as wild birds do. This shows that the response to the eye-pictures, and to eyes themselves, is an instinctive one and does not have to be re-learned in every generation.

Some caterpillars also have eye-spots, and a few tropical species have a pair which are positioned so as make the caterpillar look like a miniature snake. The scales and patterns of the snake's head are reproduced on the caterpillar's skin to enhance the effect, and, if disturbed, the caterpillar will rear up threateningly as a snake does.

The difference in size between snake and caterpillar is evidently not a problem because birds cannot estimate size very well. This is partly related to the fact that they cannot judge distance. In human beings it is the overlap of vision between the two eyes which makes it possible to tell how far away an object is. Birds, with the exception of owls and birds of prey, have their eyes on the sides of their heads and so cannot assess distance. Because of this they rely on the size of familiar objects to tell them how far away they are. To a bird, therefore, a snake-like caterpillar close to, looks like a snake a short distance away.

An astonishing variety of other animals have eye-spots. The largest and most remarkable is a toad found in Brazil which has a pair of eye-spots on its

rump. If molested it will raise its rear end in the air and point it towards the enemy who sees what looks like a large, round face with huge, all-seeing eyes.

Eye-spots are found in many other creatures including beetles, bugs and grasshoppers, and the fact that they occur so widely and in unrelated groups of animals is an indication of their useful function in defence.

Some butterflies have marks similar to eye-spots but smaller, and these are believed to have a different function to the larger ones. They are not used to scare off attackers but rather to deflect attack from the butterfly's head. These marks are generally on the hindwing at a point farthest from the head and captured butterflies very frequently have peck marks close to these spots where they have been attacked by a bird but have escaped. (Butterflies are able to fly about with part of the wing damaged or even missing.)

It is probable that a bird instinctively pecks at the head of any victim, but that small birds, not being very intelligent, have rather a limited method of deciding where the head is located. They are likely to be 'programmed' to go for a small, round, dark-coloured area at one end of the body. If the spot at the tip of the butterfly's wing fits this description better than the real head, the butterfly

Page 60: *A Bull's-eye moth from Florida; the wide distribution and diversity of eyespots is a good indication of their protective function. (NHPA, J. H. Carmichael)*

Page 61: *A snakelike caterpillar from South Africa; the difference in size between snake and caterpillar is not a problem since birds cannot judge size very well. (NHPA, A. Bannister)*

Right: *The eye-spots of the Emperor Moth are highly realistic representations of the eyes of larger animals. (NSP, P. H. Ward)*

Below: *The Peacock Butterfly shows four rather simple eye-spots on its wings. (NSP, P. H. Ward)*

Top right: *A bush-cricket from South America which is camouflaged as a dead leaf.*

Below: *A caterpillar of the Elephant Hawk-moth rears up threateningly in imitation of a snake. (NSP, P. H. Ward)*

Bottom right: *The same bush-cricket exhibiting its wings with their startling eyespots. (NSP, P. H. Ward)*

will be attacked there and may escape serious harm.

This may sound rather far-fetched but butterflies which have been found in Africa take the idea even further. They have developed distinct false heads. As well as a spot for the head they have two pairs of thin tails extending from the wing tips, one pair pointing downwards to simulate the legs and the other pointing upwardsto simulate antennae.They are even able to waggle the upper pair to make them look more like antennae. When coming into land they do a rapid about turn just as they touch down. The turn is performed so fast that it is almost imperceptible, and the tail with its false head is then correctly postioned to deceive a wating bird or lizard.

Below: *A Meadow Brown Butterfly showing its deflection spot near the wing tip; notice how inconspicuous the real head is by comparison. (NSP, P. H. Ward)*

Right: *A butterfly from Nigeria with a false head; these butterflies do a quick about-turn as they land so that the tail is where the head should be. (NSP, P. H. Ward)*

Camouflage for hunters

It is not only the hunted which can use camouflage. Hunters, too, are at an advantage if they go unseen. There are many familiar examples of basic camouflage in predators: the Tiger is striped to blend with the long grass where he stalks his prey, while the white Polar Bear and the green Praying Mantid are coloured to match the background where they lie in wait. Others are far more intricately disguised than this, and some create an illusion which will actually attract the prey, or even imitate the prey itself, so that they can approach it without causing alarm.

In the seas and rivers much of the food which fish eat consists of small worms so that predatory fish which are able to mimic these can lure smaller fish to their doom. A group which practise this way of life are the Angler Fish.

The shallow-water species of Angler Fish rest half-buried in the mud but are still able to see because their eyes are placed on the top of the head. As mentioned earlier, they are camouflaged against the sea-bed, being sandy-coloured and mottled with long, tattered membranes extending from the body.

The Angler Fish has an enormous mouth, out of all proportion to its body, which, because of its size, sucks in water with great force when opened. But its oddest feature is the fin on its back, which is very thin and elongated. This fin always flops forward, and at the end

Pages 66 & 67: *A Crab-spider lies in wait in a daisy flower. (NHPA, N. A. Callow)*

Left: *Hunters, too, can use camouflage; the Tiger's stripes break up the outline of the body, which is also disguised by countershading, the belly being lighter than the back. (NHPA, E. H. Rao)*

Below: *The Polar Bear's white fur serves as camouflage against the snow and ice where he hunts his prey. (NHPA, C. Wyatt)*

Left: *A Praying Mantid seizes a cricket which has come too close; its green camouflage makes it less obvious as it lies in wait. (Heather Angel)*

Below: *A mantid which stalks its prey on the ground and is coloured brown to match the earth; the green grasshopper has not yet noticed the mantid approaching. (NSP, P. H. Ward)*

it has a fatter, worm-like portion which hangs suspended over the mouth.

Unlike the rest of its body, the worm-like bait is conspicuous, and small fish come, unsuspectingly, to investigate. The Angler keeps quite still as they approach, but when they are close enough it suddenly flings its jaws wide open. The small fish are sucked into the Angler's cavernous mouth with the massive in-rush of water caused, and sharp, inward-pointing teeth prevent their escape.

Other Angler Fish inhabit the ocean's dark depths and hunt in a similar way except that they do not conceal themselves on the sea-bed. They do not need to hide as it is too dark for them to be seen at all, but they often have a luminous organ in the bait so that it glows to advertise its presence.

One of the crabs which has an improvised camouflage of seaweeds stuck to its shell attracts fish with its pink claws. These are the only part of the body that are not concealed, and the crab holds them aloft and sways them gently back and forth in the mass of seaweeds attached to the shell. Fish which come to see if the pink claws are edible worms find themselves seized by them and eaten by the crab.

An even stranger way of life is practised by a small fish whose colouring and pattern perfectly mimics that of the Cleaner Fish. The Cleaner Fish are themselves unusual because they eat parasites from the mouths and bodies of larger fish which permit them to approach closely and even to enter their mouths to do so.

The larger fish benefit from the attentions of the Cleaner Fish and have strong instincts against attacking them. The fish which mimic their pattern take advantage of this to come close to the

larger fish and bite a chunk out of their flesh. It is interesting that the larger fish do not pursue these Cleaner Fish mimics and take revenge. Evidently they are far too strongly inhibited by the instinct which commands them not to attack fish with such a pattern.

The treachery of the Cleaner Fish mimics is matched only by that of certain Fireflies. These exploit the system of flashing lights by which female Fireflies attract a mate. There is a sort of Morse Code to these flashes which distinguishes the females of one species from those of another, and make it easier for male Fireflies to find a female of their own kind. The females of one species, however, as well as making their own signals, will mimic the signals of another species when they feel hungry. As the hopeful males of the other species arrive they are set upon and eaten.

Mantids are a group of predatory insects which habitually lie in wait for their prey, seizing the victims with strong fore-limbs. Most have simple camouflage such as green colouring to remain inconspicuous, but some have more sophisticated disguises.

One species resembles a stick-insect and uses this resemblance to mingle with stick-insects until an opportunity arises to overpower one and eat it. Another mantid is coloured and patterned to resemble a bird-dropping, and this appearance attracts flies which are then captured.

The most impressive of all mantids, and the most remarkable of all camouflaged hunters, must be the Orchid Mantid. It is coloured a delicate shell-pink all over, even to its eyes and antennae, while the body and legs are flattened and sculptured into the curved shapes of waxy petals.

This creature is found in the jungles of Malaysia where it lurks among the blossoms of wild rhododendron bushes.

Right: *A mantid which resembles a thorny twig with a green leaf attached eats a brightly-coloured moth which it has captured with the help of its disguise. (NHPA, A. Bannister)*

Below: *This deep sea Angler Fish lures smaller fish with a fin which resembles a piece of seaweed. (A. Wheeler)*

Right: *Crab-spiders are delicate pastel shades to match the flowers, unlike the usual brownish coloration of spiders. (NSP, P. Bowman)*

Below: *A beautiful Flower Mantid, which, like the Orchid Mantid, lies in wait in flowers, and captures flying insects which come in search of nectar. (Bruce Coleman Ltd., M. Fogden)*

It is almost impossible to distinguish from the bunches of flowers, as its fore-limbs will appear to be part of one flower, its body part of another, and the hind-limbs yet another. Flying insects approach the flowers for nectar and are seized with a lightning movement by a pair of powerful, pink claws.

An imitation flower is also used as a lure by a few birds, such as the Kingbirds of North and South America. These have a bright orange crest of feathers on the crown of their head which is normally concealed by duller feathers, but can be erected and displayed, when it attracts insects to them. The bird will sit on a prominent branch and, as insects approach the 'flower', twist its neck round and snap them up. The Kingbirds do not depend entirely on this supply of food, but actively hunt insects as well.

Some much smaller animals lie in wait in real flowers, and are adequately camouflaged by being coloured to match the petals around them. The Ambush Bugs found in North America do this, and so do several species of Crab-spiders. These come in beautiful shades of orange, yellow, pink or white, quite unlike the usual brownish coloration of spiders.

Sheep in wolves' clothing

A delicate pink Crab-spider lurks among the flowers of a bramble bush and waits. An insect approaches and the spider prepares to attack. But as the insect gets closer the spider hears a buzzing sound and then notices that the insect is striped with bands of black and yellow. This spells danger to the spider, who has once before fought with a wasp and come off much the worse. The little spider backs away and allows the yellow-and-black insect to feed at the bramble flowers in peace.

What the spider does not know is that this insect is not a wasp at all but a harmless hoverfly. The hoverfly's pattern and colouring, the shape of its body and wings, even its buzzing sound, are all part of a masterly disguise. They protect it against a whole range of predators from the Crab-spider up to large mammals which would readily eat a fly but fear the sting of a wasp.

Nor are hoverflies the only insects to mimic wasps. There are several sorts of beetles with black and yellow stripes on their wing cases, slim bodies tapered at the tips and other wasp-like characteristics. One wasp-mimicking beetle will even pretend to be about to sting although beetles are incapable of this. If picked up it curls its tail under as a wasp does before stinging and sticks out a long white spine which looks like a wasp's sting but is harmless.

There are also the Clearwing Moths which are so closely modelled on wasps that they look quite unlike moths. The most unmothlike features are their thin, membranous wings, but they also have slim, stripey bodies, and heavy, wasp-like antennae. One Clearwing Moth is known as the Hornet Moth because it so closely resembles a Hornet. It is such a precise mimic that it is easily mistaken for one.

Not all hoverflies are modelled on wasps: other species resemble bees and there are also bee-mimicking moths and beetles. The beetle which mimics a bee has similar black and yellow marks on the wing-cases but is much fatter than the wasp-mimicking beetles, and is covered with fine, silvery hairs. It also mimics a bee's buzz, and from a distance

A Crab-spider attacks a fly; an equally harmless hoverfly would be protected by its resemblance to a wasp or bee. (NHPA, A. Bannister)

is quite a convincing mimic, although at close quarters it is obviously a beetle.

The bee-mimicking moths are certain hawk-moths which are only distantly related to the Clearwing Moths and it is evident that these similar disguises have evolved quite independently of each other.

It is hard to say whether wasps and bees are the most mimicked animals or whether this title should go to ants, for no-one has counted up the number of species which resemble each. Ant-mimics may well be more numerous but they are much less obvious to us since they are small and inconspicuous, and in some cases mingle with ants themselves.

Spiders are among the best-known ant-mimics although they have an immediate problem with this disguise since they have eight legs while ants (like all insects) have only six. Spiders do not have antennae, however, so they get round this problem by holding the first pair of legs high off the ground and pointing them ahead like ants' antennae. They also keep them in constant side-to-side motion to complete the effect.

Another difficulty is that an ant has two constrictions or 'waists' to its body while spiders are only naturally endowed with one. But ant-mimicking spiders have evolved an extra waist and are elongated to achieve an ant-like shape. They are coloured to match the ants, whether black or red ones, and even mimic their behaviour, as described by the naturalist Major PWG Hingston: 'The worker ants, as a rule, move quietly through the foliage, in a steady, searching kind of way. When alarmed, however, they rush about excitedly. The mimic also simulates these motions exactly. Its ordinary progression is steady and even. But the moment it is disturbed its behaviour changes. Then it assumes the bustle of the workers, simulates their

Page 80: *The Hornet Moth, with transparent wings, precisely mimics a real hornet. (NSP, J. Bradley)*

Page 81: *A hoverfly which mimics a honey-bee; the large eyes betray the fact that it is really a fly. (Heather Angel)*

Left: *A wasp-mimicing hoverfly; a real wasp has thicker antennae, smaller eyes, a narrower waist, and narrower wings held straight back along the body. (NHPA, N. A. Callow)*

Below: *A beetle which resembles a wasp. (NHPA, K. G. Preston-Mafham)*

rapid, jerky gait, perhaps darts for safety under a leaf or escapes by a sudden spring . . . Another habit of these ants is to erect the abdomen at right angles to the body and walk about with it thrust into the air. And this same peculiar attitude is sometimes assumed by the mimicking spider.'

Various insects mimic ants when they are small but soon outgrow them and have to dispense with the disguise. The young Macleay's Spectre Stick-insect for example, is about 12mm (½ inch) long when it hatches from the egg, and is dark coloured with a pinkish head, resembling a large Foraging Ant found in the same region. It even imitates the frenzied movements of this ant. As it grows, however, the dark colouring fades, becoming an inconspicuous brown before it is 20mm (¾ inch) long. It then ceases its ant-like behaviour.

Some ant-mimics such as a beetle which runs with the columns of Driver-ants found in Africa, have lost their natural shape and are elongated and constricted to give an ant-like form. But others, including certain bugs and grass-hoppers, have rectangular or rounded bodies with the outline of an ant simply 'painted-on'. The thin, curvaceous shape is reproduced in black on the very un-antlike white or green body.

Other ant disguises are only effective when viewed from above. In Central America there is a bug which has part of its hard skin behind the head greatly enlarged so that it extends over its back like a shield, completely covering the body except for the head. This shield is black and shaped like an ant, and the bug adopts the typical jerky movements of the ant. The disguise is effective, but only, of course, from above. This same jerky movement is also seen in a spider found in South America which looks like a normal spider but always carries a dead ant, larger than itself, on its back.

The Bee Beetle: its fine, silvery hairs and buzzing sound make it a convincing mimic from a distance. (NHPA, K. G. Preston-Mafham)

In this way it resembles an ant carrying a dead companion, a common sight among ants.

Why do so many insects and spiders resemble ants? As one might expect, it is to escape from predators. Ants, although small, can usually bite or sting, sometimes both, and may also spray acid in the face of an attacker. Few animals feed on them. Ants that travel in large numbers such as the Driver-ants are especially avoided.

One species of spider, however, uses its disguise to mingle with the ants and then attacks and kills them for food. Another species imitates the movements which ants make when in distress. Although these spiders are not particularly antlike in appearance, ants have poor eyesight and will respond readily to the signals for help. As they come within reach the spider devours them.

After considering the dreadful contortions which spiders, bugs, beetles and grasshoppers have gone to in order to squeeze themselves into the slim form of an ant, it seems that an insect imitating another whose basic body-plan is the same, has an easy time. Thousands of species do this, and it is the intricate detail in which colour, form and pattern are reproduced which is so impressive in these cases.

The best-known instances of this are found amongst the butterflies; the case of the Monarch and the Viceroy is a good example. The Monarch is a highly successful species. Originally native to America, it can fly long distances and now inhabits Australia, New Zealand and parts of south-east Asia. Its caterpillars feed on Milkweed plants and it is also known as the Milkweed butterfly.

It is this source of food which is the key to the Monarch's success. Milkweed plants contain substances which are

A Bee Hawk-moth feeds at a flower. (NSP, P. H. Ward)

highly poisonous to birds and mammals, but harmless to the caterpillar. As the caterpillar feeds, these substances accumulate in its body and remain in the body of the adult butterfly. Birds will eat neither the caterpillar nor the adult butterfly more than once since both make them violently sick.

Birds rapidly learn to recognize the distinctive orange and black pattern of the Monarch and to leave it alone. This is how the Viceroy benefits for its markings are very similar to those of the Monarch. Even though it is quite harmless and edible, it too is avoided by birds.

The Viceroy is found only in the Eastern states of America and is far less numerous than the Monarch. This is a condition of survival for mimics, since if they ever outnumbered the model (the poisonous species), predators would often come into contact with the edible mimic, without having had a bad experience with the model beforehand. Since the mimics are necessarily conspicuous and vulnerable, they would fast be decimated by a few, undeterred predators.

The Golden Danaid or African Monarch, a butterfly found in Asia and Australia as well as in Africa, has not one but half a dozen species of butterflies and several moths as mimics. Of the butterflies it is only the females of the various species which are mimics,

The Viceroy mimics the pattern and colouring of the Monarch and so acquires protection. (Bruce Coleman Ltd., J. Shaw)

and they take on the shape, colouring and pattern of the Golden Danaid's wings. The males have wings of the shape and pattern typical of their group of species and look quite unlike their female partners.

The advantage of this is that it halves the number of butterflies mimicking the same model. As we have already seen, it is risky for the mimic to become as numerous as the model, and with so many edible species mimicking the same model, this danger is increased.

The males are less well protected because, as with some birds, only a few are needed for the fertilization of many females, and, unlike the females who require time to produce and lay their eggs, they do not need to live very long to fulfil their useful function.

The situation is made more complex because some of these mimics also have several different colour forms for the females, each form mimicking a different distasteful species. So an individual which mimics the Golden Danaid may look quite unlike her sister who is mimicking another bad-tasting or poisonous species.

With all these confusing appearances how do the butterflies recognize each other when they need to find a mate? Some may track down their own kind by smell, but it is probable that others recognize patterns on the wings which reflect light in the ultra-violet range.

The Monarch Butterfly is poisonous and is avoided by predators such as birds. (NSP, P. H. Ward)

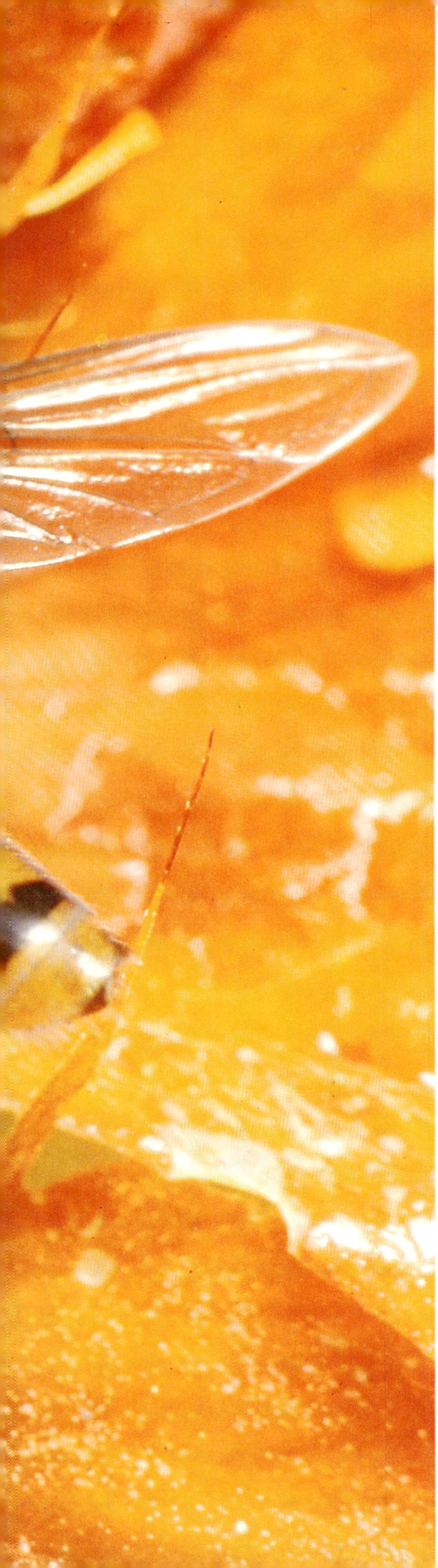

Warning colours

The Monarch and the Golden Danaid are patterned in orange and black, while their caterpillars have black and yellow stripes. But black and yellow stripes are also worn by wasps and bees. Is this just coincidence?

The answer is that it is no coincidence. The colours have come to be associated with unpleasant, distasteful or dangerous animals. All over the world creatures which have a sting or poison or simply taste bad are boldly patterned in yellow, orange or red, very often with black and sometimes with white marks on the black stripes as well. They are said to have 'warning coloration' and it is these warning colours which are mimicked by the edible imposters.

The advantage in this international convention is that it becomes recognized by predators. If each distasteful species had quite individual colours every single predator would have to try at least one of that species to learn that it should be avoided. Many individuals would be killed just to teach the predator a lesson. By wearing a recognized uniform, a sort of badge of nastiness, they save both themselves and the predators a lot of trouble.

Any unpleasant species may have its own idiosyncratic warning colours. In Africa two species of aggressive and bad-tasting birds called Drongos are coloured jet-black, and two other edible species mimic their colouring and acquire protection. Their warning colours must be effective or they would not have survived, but on the whole unconventional warning colours are at a disadvantage. They may succeed, but the species which has donned the more usual red and black or yellow and black will do better.

Warning colours of yellow, orange or red with black stripes are worn by a vast number of animals which are dangerous or bad-tasting, and also by harmless and edible imposters, in this case a hoverfly. (NHPA, N. A. Callow)

Cuckoos and Bee Orchids

An interesting sort of mimicry is found in cuckoos: not in the birds themselves, but in their eggs. As is well known many cuckoos lay their eggs in the nests of other birds which then rear the baby cuckoo as if it were their own. Not all cuckoo species do this. Some build a nest and raise their own young, and these provide an interesting comparison with the parasitic cuckoos.

The non-parasitic species usually lay quite large, plain, white eggs. The parasitic species almost all lay smaller eggs, and these are coloured and patterned to mimic the eggs of the birds who unwittingly adopt them. Many different birds are used as foster-parents by cuckoos, but generally, in any one region, a cuckoo species parasitizes one particular bird, and the cuckoo's eggs specifically imitate the eggs of that bird.

If the nest belongs to a large, intelligent species, such as a Crow or a Magpie, the egg mimicry is very good. Smaller birds are not so intelligent, and can be duped just as easily by an egg which is only an approximate copy of their own. Most can detect that an egg is an alien one if the pattern and colour are very different, and will eject it from the nest. But they do not seem perturbed by an egg which is larger than their own. Often the pattern and colouring are the same but the cuckoo's eggs are half as large again as those of the foster parent.

It may even be that birds prefer larger eggs. Experiments with Herring Gulls (which, incidentally, are not parasitized by cuckoos), have shown that this is so for them. If a model of the Herring Gull's egg, correct in colour, shape and pattern but three times larger than normal, is offered to a female gull, she will brood it in preference to her own. The egg is acting as a 'super-stimulus'. The large, fake egg compels the gull to incubate it far more strongly than one of her own, much smaller eggs.

The same mechanism is at work when small birds which have hatched a cuckoo's egg continue to feed the nestling even though it grows to twice their size. A tiny Reed Warbler will persist in feeding even though it may

Left: *The flower of the Bee Orchid mimics a female bee. (NHPA, S. Dalton)*

Page 90: *Hairs on the lip of the flower feel like those on the body of the female bee and stimulate the male bee to try to mate with the flower. (NSP, P. A. Bowman)*

Page 91: *This orchid, known as a Fly Orchid, is in fact pollinated by a small burrowing wasp, whose female the flower resembles. (NSP, M. Chinery)*

have to perch on the back of the grotesquely large cuckoo chick in order to reach its beak.

So powerful is the sight of the cuckoo chick's open beak that another small bird passing by will respond to it, although it has never seen the chick before. The passing bird will suddenly stop and push into the greedy mouth a morsel of food which was destined for its own young ones.

A cuckoo which is fostered by the more intelligent Crows and Magpies needs to be more restrained. It has to share the nest with the foster bird's own young (those that parasitize small birds just push the other eggs out of the nest as soon as they can). But since these larger birds are less gullible the cuckoo chick needs to mimic one of their own nestlings in order to be fed. Such cuckoo chicks have evolved colouring on the tops of their heads, on their backs and inside their beaks which mimics that of the other young birds. Their undersides, however, look quite different, but this does not matter since the parent bird never sees them from below.

So many animals, especially insects, imitate parts of plants (leaves, twigs, stems or flowers) that it might seem as though plants had all the advantages and there was nothing enviable at all about being an insect. But there are, in fact, a few plants which have proved the reverse and benefit by mimicking insects. These are the Bee Orchids.

The great disadvantage which plants have is that, unlike animals, they are immobile, and this is a problem when they need to be fertilized. The solution to this problem is pollen: tiny, light-weight particles containing male cells, which are blown by the wind or carried by insects to fertilize the flowers of other plants of the same species.

As is well known, flowers usually attract insects by having nectar which the insects eat and by being brightly coloured and strongly scented to advertise this supply of food. When the insects visit the flowers to collect the nectar they inadvertently pick up pollen. They later transfer it to another flower, again by chance, and this is known as 'pollination'.

Bee Orchids have a different, and rather exotic, way of persuading insects to pollinate them. Their flowers resemble female bees to such a degree that male bees are attracted to them and will try to mate with the flower. After an unsuccessful attempt to mate, the male bee, probably feeling rather confused and frustrated, goes off to try his luck elsewhere. Should he try to mate with another flower he will, in the process, transfer pollen from the first flower to the second.

The flowers of these orchids not only resemble the bees visually but also produce scents very similar to those the female bee produces, and have hairs on the lip of the flower which feel like those on the body of the female. With real bees these hairs are known to play an important role in stimulating the male to mate with the female.

Each species of orchid mimics only one particular species of bee, and no other bees will try to mate with them. This is beneficial to the orchids because they need to be pollinated by pollen from one of their own kind and no other.

Other orchids mimic female wasps to get themselves pollinated, for example the Tartan Tongue Orchid of Australia whose flowers resemble the females of a certain Ichneumon Wasp. One strange orchid found in Australia seems to imitate an ant. The lip of the flower has shining, globular, black growths which are grouped together in the shape of an ant. The lip of the Ant Orchid, as it is called, quivers in the wind, making the 'ant' seem animated. This odd structure is so striking in its resemblance to an ant that it has been suggested it must have a special function in pollination, but no-one has yet proved that this is so. Like many other strange appearances and interesting resemblances in animals and plants, it is a mystery which waits to be investigated and which promises to yield another fascinating story of nature's camouflage.

Left: *A cuckoo's egg in the nest of a Meadow Pipit; the egg is only an approximate copy of the foster bird's own since small birds are easily duped. (NSP, T. D. Bonsah)*

Endpapers: *A pair of Red-capped Robins; the male is brightly coloured while the female has dull feathers so that she will be camouflaged when sitting on the nest. (NHPA, H. Newman)*